The Role of Livestock in Developing Countries

Ridwan Mohamed

Bibliographic information published by the German National Library:

The German National Library lists this publication in the National Bibliography; detailed bibliographic data are available on the Internet at http://dnb.dnb.de.

ISBN: 9783346953544
This book is also available as an ebook.

© GRIN Publishing GmbH
Trappentreustraße 1
80339 München

Print and binding: Books on Demand GmbH, Norderstedt, Germany
Printed on acid-free paper from responsible sources.

GRIN web shop: https://www.grin.com/document/1400268

Review Date: 7th September, 2022

Review of Role of Livestock in Developing in Developing Countries

Ridwan Osman Mohamed[1]

1. College of Agriculture and Veterinary Medicine
2. Department of Agricultural Economics and Agribusiness Management; Jimma University, Ethiopia.

Table of Contents

List of Tables

Abstract

Livestock play a significant role in rural livelihoods and the economies of developing countries. They are providers of income and employment for producers and others working in, sometimes complex, value chains. They are a crucial asset and safety net for the poor, especially for women and pastoralist groups, and they provide an important source of nourishment for billions of rural and urban households. These socio-economic roles and others are increasing in importance as the sector grows because of increasing human populations, incomes and urbanisation rates. To provide these benefits, the sector uses a significant amount of land, water, biomass and other resources and emits a considerable quantity of greenhouse gases. There is concern on how to manage the sector's growth, so that these benefits can be attained at a lower environmental cost. Livestock and environment interactions in developing countries in are both positive and negative. Livestock production in the developing world occurs in a wide range of heterogeneous production systems. These can range from pastoral/grassland-based systems, which occupy most of the land area and have low human population densities, through mixed crop-livestock systems, usually in areas suitable both for arable and livestock production and where the bulk of rural human population lives, and intensive systems usually in peri-urban/urban areas. New diversification options and improved safety nets will also be essential when intensification is not the primary avenue for developing the livestock sector. These processes will need to be supported by agile and effective public and private institutions.

Key words: livestock, developing countries, economics importance

1. Introduction

Livestock plays a crucial role in developing countries included Somaliland. Among rural communities in developing countries, livestock still serves as the primary exchange market where sheep, camels and cows are bartered. With increased investment, smart regulation, infrastructure development and sector coordination, developing countries have ample opportunities to capitalize on the growth of the livestock sector in the developed and developing countries, while supporting its own growing local consumption.

We are at a moment in time where our actions could be decisive for the resilience of the world food system, the environment and a billion poor people in the developing world, let alone for the fate of our planet. The society has realised that there are significant pressures on the world's food and ecological systems, where the alterations of global biogeochemical cycles could be irreversible and where new drivers, such as climate change, are likely to exert additional pressures for sustainably feeding 9 billion people in the future. At the same time, and especially in the developing world, the demand for livestock products is increasing, thus adding additional pressures on the world natural resources.

It is essential to dissect the discussion on the roles of livestock, as the economic development of different countries, their structure of production, the demand for livestock products, the competition with other sectors and others shape these roles, making broad generalisations about the livestock sector useless (and dangerous) for informing the current global debates on food security and the environment. It is essential to deliver nuanced, scientifically informed messages about livestock's roles in relation to food systems, livelihoods and their economic and environmental performance.

Livestock are an integral component of agriculture in developing countries and make multifaceted contributions to the growth and development of the agricultural sector. Livestock help improve food and nutritional security by providing nutrient-rich food products, generate income and employment and act as a cushion against crop failure, provide draught power and manure inputs to the crop subsector, and contribute to foreign exchange through exports. Also, by using crop residues as feed, livestock save land for food production that would otherwise be used for fodder production. Additionally, livestock make substantial contributions to environmental conservation, supplying draught power and manure for fertilizer and domestic fuel that save on the use of petro-products.

This paper reviews the economic roles of livestock in the developing countries. It also discusses key factors that are likely to determine the future contribution of the sector to food security, environmental protection and economic growth. The paper proposes key actions for improving different aspects of livestock systems so that the positive roles outweigh the negatives.

2. Literature Body

2.1 Export trade in meat and live animals

Despite this recent expansion, the value of the official trade in meat and live animals is still only a fraction of the value of unofficial cross-border live animal exports. Since it is considered to be illegal by the Ethiopian authorities, the unofficial cross border trade is poorly documented, and its exact value is unknown. Estimating the extent of the unofficial cross border trade from Ethiopia is further complicated because animals move along multiple routes out of Ethiopia into four adjoining countries – Sudan, Somalia/Somaliland/Puntland, Kenya and Djibouti. Unfortunately, estimates of the scale of these movements refer to different time periods for each route:

- Somaliland: $42 million from Somali Region, circa 2003 (Nin Pratt et al. 2005); $57 million in 2009 (Little et al. 2010, citing Somaliland Chamber of Commerce)
- Puntland: The Puntland port of Bosasso became a major competitor to the port of Berbera in Somaliland after 2000 following the imposition of a ban on livestock imports imposed by Saudi Arabia due to concerns about animal health. By 2008-09, exports from Berbera and Bosasso were roughly equal at about 1.2 million head of sheep and goats and 75,000 head of cattle annually. Very approximately, exports via Puntland may be of equivalent value to those from Somaliland - $57 million in 2009 (Little et al. 2010, citing Somaliland Chamber of Commerce)
- Sudan via North Gondar: $18 million, circa 2007 (Mulugeta et al. 2007)
- Kenya via Moyale: $11 million from southern Ethiopia in 2001 (Mahmoud 2003, cited in Mahmoud 2010); $10 million from Somali Region, circa 2003 (Nin Pratt et al. 2005)

- Djibouti: Became a major player in livestock exports when it gained privileged access to the Saudi market in late 2006 after opening a Regional Quarantine Facility to certify the health of exported animals. At that time competing ports in Somalia and Somaliland were banned officially from shipping to the Saudi market due to fears about Rift Valley Fever. 1.5 million sheep and goats were exported from Djibouti in 2007 and 2008, falling to 1 million in 2009 after the RVF ban was lifted (Majid 2010). We have no information on the proportion of the animals that transited legally or illegally from Ethiopia into Djibouti, or their monetary value.

- The Livestock Marketing Authority estimated total unofficial exports on all routes except to Sudan at US $105-107 million, circa 2000 (Hurissa and Eshetu 2002; LDMPS – Phase I, Vol. O, pp 21 citing the Livestock Marketing Authority). More recently, the Ethiopian Sanitary and Phytosanitary Standards and Livestock and Meat Marketing Program (SPSLMM) estimated the value of the unofficial cross border trade at between $250 and $300 million US dollars per year (Little et al. 2010).

The bar to the right of Figure 2 compares official live animal and meat exports for 2008-09 with estimates of the different unofficial cross border trade flows enumerated above. About 50% of the sheep and goats exported through Berbera originate in Ethiopia's Somali Region, and it is likely that a similar proportion of the animals exported from Bosasso come from Ethiopia (Majid 2010, citing Holleman 2002).

2.2 Livestock population and production in different agro-ecological zones

1.5 billion head of cattle and buffaloes, and the 1.7 billion sheep and goats are fairly evenly distributed across the land-based systems, but average densities increase sharply from grazing systems to mixed irrigated systems; the latter have far greater livestock supporting capacities per unit area (); And further, about 87.1% (32.7 million) of the world's camel population is in Africa, with 13.71 million camels in Eastern African countries, including Kenya, Somalia, Ethiopia, and South Sudan (FAOSTAT, 2019).. Only a small fraction of the world's ruminant population is found in industrial feedlots, partly because this corresponds to only the final stage of the animal's life cycle, even in regions where intensive production is common. Ruminant feedlots are predominantly a North American phenomenon, though they are used to a lesser extent in parts of Europe and the Near East. The vast majority of large and small ruminant populations are found in the developing regions: some 70% of small ruminants in grazing systems and over 80% of large ruminants in grazing systems are located in developing regions. These shares are respectively about 80% and 70% in rainfed mixed systems and 87% and 92% in irrigated mixed systems (H. Steinfeld, *et al*, 2006).

Ruminant productivity varies considerably within each system, but in grazing and mixed systems overall productivity is lower in developing countries than in developed ones: in grazing systems, for example, worldwide beef production per head averages 36 kg/head/year, but the average for developing countries is only 29 kg/head/year. In the mixed rainfed system, the difference between developed and developing regions is even more marked. By far the largest variation in intensity of production is found within this system, which is the largest producer of ruminant products. Even though the developing regions host the vast majority of the mixed rainfed ruminant population, they account for less than half of the system's production worldwide. In fact beef productivity in these regions averages 26 kg/head, as opposed to 46 kg/head at world level, and their milk production represents only 22% of the world total. Across all systems, developing regions account for half of the world's beef production, some 70% of mutton production and about 40% of milk production. A sharply contrasting situation is found in the monogastric sector. More than half of the world's pork production currently originates from industrial systems, and over 70% of poultry meat (H. Steinfeld, *et al*, 2006).

About half of this production originates from developing countries and, though reliable population figures are not available, variation in productivity between regions is probably much lower than for ruminants. Including the substantial monogastric production from irrigated mixed systems in developing regions, these regions account for the majority of the world's pork, poultry and egg production. Huge differences are found between the developing regions: although substantial, total production in Latin America is less than one tenth of that in Asia, and production in Africa and the Near East is almost non-existent. The developed countries and Asia together account for over 95% of the world's industrial pork production (H. Steinfeld, *et al*, 2006).

Table 1: Livestock population and production in different agro-ecological zones (2001-2003)

Type of animal / animal product	Livestock population (10^6 heads) and production (10^6 tonnes)		
	Arid and semi-arid tropics and and sub-tropics	Humid and sub-humid tropics and sub-tropics	Temperate and tropical highlands
Animal			
Cattle and buffaloes	515	603	381
Sheep and goats	810	405	552
Animal product			
Total beef	11.7	18.1	27.1
Total mutton	4.5	2.3	5.1
Total pork	4.7	19.4	18.4
Total poultry meat	4.2	8.1	8.6
Total milk	177.2	73.6	343.5
Total eggs	4.65	10.2	8.3

Source: based on FAOSTAT data and calculations by J. Groenewold ('Classification and characterization of world livestock production systems'; unpublished report for the Food and Agriculture Organization, 2005)

2.3 Livestock Production, Economics and Trade

Livestock production in the developing world occurs in a wide range of heterogeneous production systems. These can range from pastoral/grassland-based systems, which occupy most of the land area and have low human population densities, through mixed crop-livestock systems, usually in areas suitable both for arable and livestock production and where the bulk of rural human population lives, and intensive systems usually in peri-urban/urban areas. Landless systems are also often found in urban areas. All these systems in developing countries produced about 50% of the beef, 41% of the milk, 72% of the lamb, 59% of the pork and 53% of the poultry, globally (Herrero et al., 2009). These shares are likely to increase, as most future growth in livestock production is projected to occur in the developing world (Bruinsma, 2003; Rosegrant et al,. 2009). Most meat and milk in the developing world comes from mixed systems (Sere´ and Steinfeld, 1996; Steinfeld et al., 2006; Herrero et al., 2009). These systems play a very important role in global food security, as they also produce close to 50% of the global cereal output (Herrero et al., 2009 and 2010). However, the highest rates of increase in animal production observed in the last decades, and forecasted into the future, are in the intensive pig and poultry sectors of the developing world (Delgado et al., 1999; Bruinsma, 2003; Steinfeld et al., 2006).

Livestock production in the developing world is also an important economic activity. Livestock products are high-value products, especially when compared with crops. For example, the average global price of a tonne of red meat is more than 10 times higher than the price of soya bean, whereas that of milk is 70% higher (data from FAOSTAT, 2011). This makes milk and meat to rank as some of the agricultural commodities with the highest gross value of production (VOP) in the developing world (FAOSTAT, 2011). In the last decade, livestock have represented between 17% and 47% of the total agricultural VOP in developing country regions (range defined by South East (SE) Asia and Central America, respectively; FAOSTAT, 2011). Over the last 40 years, the value of livestock production has seen an average 2.7% growth per year in sub-Saharan Africa (SSA), 3.4% in Central America and 4.1% in SE Asia. These indicators of growth compare favourably with, for example, a mean annual growth in VOP of 1.2% in North America over the same period (FAOSTAT, 2011). These growth rates are largely a reflection of increased production in the developing world.

Trade is an important dimension in the economics of livestock. Local consumption dominates livestock product demand, and international trade is relatively small. However, international trade has increased in recent years as a result of trade liberalisation and lack of competitiveness and low technological change in some regions (Table 1; FAO (Food and Agricultural Organization), 2009). Dairy and eggs dominate trade, but meat exports are important for a handful of countries (i.e. Brazil, Thailand; FAO, 2009). Most trade of livestock products occurs within a country, with movements of animal products, inputs and services being very dynamic because of increased internal connectivity, transport networks, improved value chains and the increasing need to supply the growing urban populations.

2.4 The value of livestock as credit

Small ruminant livestock plays multiple roles (besides financial benefits from meat sales) in the wellbeing of poor and landless households in arid regions of sub-Saharan Africa (Bettencourt et al. 2015). First, for many farm households in rural economies, sheep and goat serve as a major form of savings and investment as well as financial security against deficits in household earnings. The animals also assume insurance role in rural livelihoods to overcome unforeseen necessities, including settling of medical bills and school fees (Verpoorten 2009). In rural Africa, financial markets are non-existent, or even if available, smallholder farmers face serious limitations in accessing services from them (Islam and Maitra 2012). In the absence of strong financial markets, livestock, including sheep and goats, is used as alternative forms of wealth accumulation (savings or financing) and risk-coping strategy (insurance) (Islam and Maitra 2012). The animals are, therefore, used as both short- and long-term savings against future needs.

In practice this calculation is not straightforward. Rural financial markets are often described as fragmented in the sense that different kinds of borrowers are serviced by a wide array of lenders offering loans subject to different repayment conditions. As a result, the interest rates changed by rural lenders can vary widely within a community, ranging from zero for loans between kin and friends to well over 100% per annum on loans from professional moneylenders (Banerjee 2003, cited in Conning and Udry 2005). Nominal interest rates are also misleading when high rates of monetary inflation reduce the real interest rate that borrowers actually pay.

2.5 Transport and haulage by equines

This section estimates the national economic benefits derived from the use of equines – horses, donkeys and mules – as providers of transport, traction and haulage. Ethiopia is home to a lot of equines. According to FAOSTATS, Ethiopia contains over 40% of sub- Saharan Africa's horses and donkeys and over 90% of the subcontinent's mules. World-wide, only China has more donkeys than Ethiopia (FAOSTAT http://www.fao.org/corp/statistics/). Ethiopian equines are working animals. Pack and riding animals compete successfully with wheeled vehicles because of the country's rugged terrain and poor road network, both in remote rural areas and in parts of some cities. Equine power therefore has a clear economic value, which is implicitly recognized by MOFED since it assigns producer prices to equines. However, MOFED has not developed output coefficients to estimate equine work capacities, and neither CSA nor MOFED assign producer prices to the services supplied by equines. Aside from their value as stock or as a store of value, for purposes of calculating agricultural GDP, equines might as well not exist.[4]

2.6 Livestock production development in economic

Although livestock ownership is often seen as a sign of wealth – household typically move up the 'livestock ladder' from poultry to goats or sheep, to cattle/buffalo (Dercon and Krishnan, 1996; Ellis and Freeman, 2004; Kristjanson et al., 2005; Deshingkar et al., 2008) – livestock's share of income was highest in the poorest income quintile, which shows that they are important to the poor as well (Davis et al., 2007). The growth in demand for milk and meat, mainly driven by urban consumers in developing countries, has been increasing in the last few decades and is projected to double by 2050 (Delgado et al., 1999; Rosegrant et al., 2009). This rising demand for milk, meat, fish and eggs has generated jobs all along the livestock value chain, from input sales through animal production, trading and processing to retail sales.

2.7 Livestock in job creation and financial asset

Trading and processing jobs in the livestock sector are especially high in the informal sectors of countries in Asia and Africa, where most meat, milk, eggs and fish are sold (Grace et al., 2008) and where most of the people selling and buying livestock foods are themselves poor (Omore et al., 2001; Kaitibie et al., 2008). Street food is a large part of the informal sector in most developing countries – the largest in South Africa (Perry and Grace, 2009) – and therefore a major source of income and employment for the poor. Animal source foods are among the most commonly sold street foods (Perry and Grace, 2009), and it is poor women who do most of the work preparing and selling these foods. It is estimated that up to 1.3 billion people globally are employed in different livestock product value chains globally (Herrero et al., 2009).

Livestock are often one of the main assets that rural households possess. Access to, control over and ownership of assets are critical aspects of well-being (Sherraden, 1991; Carter and Barrett, 2006). Assets are stores of wealth that can be sold to finance investments such as school fees or in time of need such as an illness or drought. Assets can act as collateral and facilitate access to credit and financial services, as well as increase social status. In their study of 'voices of the poor', Narayan et al. (2000) found that 'the poor rarely speak of income, but focus instead on managing assets – physical, human, social and environmental – as a way to cope with their vulnerability'.

2.8 Livestock manure as agricultural natural fertilizers

Across the developing world, people keep livestock to earn income, increase their crop production (animals provide manure for fertilising crop fields and traction for ploughing and transporting goods to markets), store wealth and to feed their families. In the absence of banks and insurance policies, livestock serve as 'piggy banks', a way for people to save and store money and manage risk. Despite their multiple benefits, livestock are also associated with negative impacts on health (though food safety or zoonotic disease) and on the environment. Helping household to increase the benefits and minimise the risks associated with livestock can make an important contribution to improving livelihoods and reducing poverty.

3. Discussion findings

1.5 billion head of cattle and buffaloes, and the 1.7 billion sheep and goats and about 87.1% (32.7 million) of the world's camel population is in Africa which is nearly 1005 of the camels are reared in the developing countries which are the economic important in those countries. The economic importance of livestock and livestock products in the developing countries is approving that the price of meat is 10 times higher than soya bean which is still is higher demanded according to the counterpart of soya bean in addition to milk demand of 70% higher than soya bean. The credit or financing benefits of livestock derive from the ability of livestock owners to dispose of their animals for particular purposes at a time of needs as an emergent economy sac, credit source and financial bank in a household based of the pastoral rural communities. Livestock are used as slaughtering in household level, inviting guest and religious sacrifices as well as traditional meeting spiritualities.

Livestock is industry of job creation when livestock owners are hiring the employee for livestock health services, livestock keepers, and other farm related activities. Trading and processing jobs in the livestock sector are especially high in the informal sectors, where most meat, milk, eggs and fish are sold and where most of the people selling and buying livestock foods are themselves poor which is making livestock to be backbone of the economy in the developing countries. Livestock manure is natural fertilizer whereas livestock farmers use for their agricultural crops to fertilize crop farmers as well as sold sometimes although there is need for further investments in this sector of livestock fertilizers to develop the natural production of fertilizers in the developing countries relating with its policy engagement.

For further in those findings there is need a researches to induce the whole intended logical framework in which developing countries can combine the different production livestock sectors to develop their economy in livestock context.

4. Conclusion and Discussion

Livestock play a significant role in rural livelihoods and the economies of developing countries. They are providers of income and employment for producers and others working in, sometimes complex, value chains. In the most of the developing countries, people keep livestock to earn income, increase their crop production (animals provide manure for fertilising crop fields and traction for ploughing and transporting goods to markets), store wealth and to feed their families. The analyses presented here have demonstrated the complex balancing act of weighing the roles that livestock play in the developing countries economy. On the one hand, we acknowledge that livestock is an important contributor to the economies of developing nations, to the incomes and livelihoods of millions of poor and vulnerable producers and consumers, and it is an important source of nourishment.

This research review has indentified that livestock are often one of the main assets that rural households possess. However, the debate needs to increase in sophistication so that the poor and undernourished are not the victims of generalisations that may translate into policies or reduced support for the livestock sector in parts of the developing countries where the multiple benefits of livestock outweigh the problems it causes. The balance between these ways of farming is likely to be different in different regions of the world, but understanding where the balance should lie for achieving socially and environmentally goals is still the question that merits significant research.

Hence, investment in developing efficient value chains (including market development, service provision, adequate institutional support, etc.) should be high in the development agenda, to create well exercisable manner livestock economy with government support to develop the national economies of their respective countries.

Lastly this research review is recommending developing the livestock sectors in the developing countries to support over one billion poor livestock beneficiaries in the world. There is also need for further research studies in the developing countries's economy to emphasis sector of economic importance in that country.

5. Reference

1. Bruinsma J 2003. World agriculture: towards 2015/2030 – an FAO perspective. Earthscan Publications, London, UK.
2. Herrero M, Thornton PK, Gerber P and Reid RS 2009. Livestock, livelihoods and the environment: understanding the trade-offs. Current Opinion in Environmental Sustainability 1, 111–120.
3. Herrero M, Thornton PK, Havlı́k P and Rufino M 2011. Livestock and greenhouse gas emissions: mitigation options and trade-offs. In Climate change mitigation and agriculture (ed. E Wollenberg, A Nihart, ML Tapio-Bistrom and C Seeberg- Elverfeldt) Earthscan, London, UK (in press).
4. Herrero M, Thornton PK, Notenbaert AM, Wood S, Msangi S, Freeman HA, Bossio D, Dixon J, Peters M, van de Steeg J, Lynam J, Parthasarathy Rao P, Macmillan S, Gerard B, McDermott J, Sere´ C and Rosegrant M 2010. Smart investments in sustainable food production: revisiting mixed crop-livestock systems. Science 327, 822–825.
5. Delgado C, Rosegrant M, Steinfeld H, Ehui S and Courbois C 1999. Livestock to 2020: the next food revolution. Food, Agriculture and the Environment Discussion Paper 28. IFPRI/FAO/ILRI, Washington, DC, USA.
6. Steinfeld H, Gerber P, Wassenaar T, Castel V, Rosales M and de Haas C 2006. Livestock's long shadow. Environmental issues and options. LEAD-FAO, Food and Agriculture Organization, Rome, Italy, 390pp.
7. Rosegrant MW, Fernandez M, Sinha A, Alder J, Ahammad H, de Fraiture C, Eickhout B, Fonseca J, Huang J, Koyama O, Omezzine AM, Pingali P, Ramirez R, Ringler C, Robinson S, Thornton P, van Vuuren D, Yana-Shapiro H, Ebi K, Kruska R, Munjal P, Narrod C, Ray S, Sulser T, Tamagno C, van Oorschot M and Zhu T 2009. Looking into the future for agriculture and AKST (Agricultural Knowledge Science and Technology). In Agriculture at a crossroads (ed. BD McIntyre, HR Herren, J Wakhungu, RT Watson), pp. 307–376. Island Press, Washington, DC.
8. FAO 2007. State of Food and Agriculture Report: paying farmers for environmental services. Food Agricultural Organisation, United Nations. http://www.fao.org/docrep/010/a1200e/a1200e00.htm
9. FAO (Food and Agriculture Organization) 2009. The State of Food and Agriculture. Livestock in the balance. Food and Agriculture Organization of the United Nations, Rome, Italy.
10. FAO 2010. Greenhouse gas emissions from the dairy sector. A life cycle assessment. Food and Agriculture Organization of the United Nations, Rome, Italy FAO 2009.
11. FAOSTAT 2011. FAOSTAT database. FAO, Rome, Italy
12. Bettencourt EMV, Tilman M, Narciso V, Carvalho MLS, Henriques PDS (2015) The livestock roles in the wellbeing of rural communities of Timor-Leste. RESR, Piracicaba-SP 53(1):S063–S080. https://doi.org/10.1590/1234-56781806-94790053s01005
13. Food and Agriculture Organization Corporate Statistical Database (FAOSTAT), FAOSTAT, 2019, http://www.fao.org/faostat/en/#data/QA.
14. Islam A, Maitra P (2012) Health shocks and consumption smoothing in rural households: does microcredit have a role to play? J Dev Econ 97(2):232–243. https://doi.org/10.1016/j.jdeveco.2011.05.003

15. Verpoorten M (2009) Household coping in war- and peacetime: cattle sales in Rwanda, 1991–2001. J Dev Econ 88(1):67–86. https://doi.org/10.1016/j.jdeveco.2008.01.003

16. Davis B, Winters P, Carletto G, Covarrubias K, Quinones E, Zezza A, Stamoulis K, Bonomi G and DiGiuseppe S 2007. Rural income generating activities: a cross country comparison. ESA Working Paper 07-16. FAO, Rome.

17. Dercon S and Krishnan T 1996. Income portfolio in rural Ethiopia and Tanzania: choices and constraints. The Journal of Development Studies 32, 850–875.

18. Deshingkar P, Farrington J, Rao L, Sharma SAP, Freeman A and Reddy J 2008. Livestock and poverty reduction in India: findings from the ODI Livelihoods Options Project. Discussion Paper No. 8. Targeting and Innovation. ILRI (International Livestock Research Institute), Nairobi, Kenya.

19. Ellis F and Freeman HA 2004. Rural livelihoods and poverty reduction strategies in four African countries. The Journal of Development Studies 40, 1–30.

20. Kristjanson P, Krishna A, Radeny M, Kuan J, Quilca G and Sanchez-Urrelo A 2005. Dynamic Poverty Processes and the Role of Livestock in Peru. FAO/Pro- Poor Livestock Policy Initiative Working Paper.

21. Kristjanson P, Waters-Bayer A, Johnson N, Tipilda A, Jemimah N, Batenwreck I, Grace D and MacMillan S 2010. Livestock and women's livelihoods: a review of the recent evidence. Discussion Paper 20. ILRI, Addis Ababa, Ethiopia.

22. Kaitibie S, Omore A, Rick K, Salasya B, Hooton N, Mwero D and Kristjanson P 2008. Influence pathways and economic impacts of policy change in the Kenyan dairy sector, ILRI Research Report No. 15. ILRI, Nairobi, Kenya, 40p.

23. Omore A, Cheng'ole MJ, Fakhrul ISM, Nurah G, Khan MI, Staal SJ and Dugdill BT 2001. Employment generation through small-scale dairy marketing and processing: experiences from Kenya, Bangladesh and Ghana, ILRI (http: www.fao.org/ag/againfo/resources/en/pubs_aprod.html#1).

24. Omore A, Arimi S, Kangethe E, McDermott J, Staal S, Ouma E, Odhiambo J, Mwangi A, Aboge G, Koroti E and Koech R 2001. Assessing and managing milkborne health risks for the benefit of consumers in Kenya. SDP research report. International Livestock Research Institute, Nairobi, Kenya.

25. Perry B and Grace D 2009. The impacts of livestock diseases and their control on growth and development processes that are pro-poor. Philosophical Transactions of the Royal Society B: Biological Sciences 364, 2643–2655.

26. Grace D 2007. Women's reliance on livestock in developing-country cities. ILRI Working Paper. International Livestock Research Institute, Nairobi, Kenya.

27. Grace D, Randolph T, Olawoye J, Dipelou M and Kang'ethe E 2008. Participatory risk assessment: a new approach for safer food in vulnerable African communities. Development in Practice 18, 611–618.

28. Grace D, Mutua F, Ochungo P, Kruska R, Jones K, Brierley L, Lapar L, Said M, Herrero M, Pham Duc P, Nguyen BT, Akuku I and Ogutu F 2012. Mapping of poverty and likely zoonoses hotspots: report to the Department for International Development. ILRI report International Livestock Research Institute, Nairobi, Kenya.

29. Sherraden M 1991. Assets and the poor: a new American welfare policy. ME Sharpe, Armonk, NY.

30. Carter MR and Barrett CB 2006. The economics of poverty traps and persistent poverty: an asset-based approach. Journal of Development Studies 42, 179–199.

31. Narayan D, Patel R, Schafft K, Rademacher A and Koch-Schulte S 2000. Voices of the Poor. Can Anyone Hear Us? Voices from 46 Countries. The World Bank, Washington, DC.

32. H. Steinfeld, T. Wassenaar & S. Jutzi. 2006. Livestock *production systems in developing countries: status, drivers, trends*. Global drivers of the livestock sector. Rev. sci. tech. Off. int. Epiz., 2006,=25 (2), 505-516.